FAITH AND PHYSICS UNCOVERED

The Compelling Relationship Between
Timeless Faith and Modern Physics

RODGER PRICE

Faith and Physics Uncovered
The Compelling Relationship Between Timeless Faith and Modern Physics
Rodger Price © 2022

The author and the publisher support copyright and the sharing of thoughts and ideas it enables. Thank you for buying an authorized edition of this book and honoring our request to obtain permission for any use of any part of this publication, whether reproduced, transmitted in any form or by any means, electronic, mechanical, photocopying, recording, or otherwise, or stored in a retrieval system. Your honorable actions support all writers and allow the publishing industry to continue publishing books for all readers. All rights reserved.

While the publisher and author have used their best efforts in preparing this book, they make no representations or warranties with respect to the accuracy or completeness of this book and specifically disclaim any implied warranties of merchantability or fitness for a particular purpose. No warranty may be created or extended by sales representatives or written sales materials. Neither the publisher nor the author shall be liable for any loss of profit or any other commercial damages, including but not limited to special, incidental, consequential, or other damages. The stories and interviews in this book are true although the names and identifiable information may have been changed to maintain confidentiality.

The publisher and author shall have neither liability nor responsibility to any person or entity with respect to loss, damage, or injury caused or alleged to be caused directly or indirectly by the information contained in this book.

Bible references from the New International Version (NIV) unless otherwise specified.

Softcover ISBN: 978-1-61206-253-2
Hardcover ISBN: 978-1-61206-254-9

To purchase this book at quantity discounts, contact Aloha Publishing at alohapublishing@gmail.com

AlohaPublishing.com

Printed in the United States of America

The writing of this book would not have happened but for my brother Ron. He encouraged me, 25 years ago, to write down my experiences in my spiritual journey and had started to teach to others. He also paved the path for me to publish what I wrote. Thank you, Ron!

And of course, none of this would have happened without the amazing grace of God—the Great I AM! Thank you, Lord, for how you bless me in so many ways.

CONTENTS

Introduction	9
Chapter 1: What Is Faith?	13
Chapter 2: The Size of Things	19
Chapter 3: The Light of the World	33
Chapter 4: The Illusion of Physical Reality	59
Chapter 5: What It All Means	81
Acknowledgments	91
About the Author	93
Connect With Me	95

INTRODUCTION

This book is about the world as seen through the principles of physics and how it impacted my spiritual journey. I was a mechanical engineer in my early 30s when a friend suggested I read the book *In Search of Schrödinger's Cat: Quantum Physics and Reality* by John Gribbin. Reading this book rocked my world.

In search of Schrödinger's Cat is about the history of physics and has nothing to do with God, but it took me on a path that led right to God's doorstep. Reading about what scientists have learned about our world—in the twentieth century in particular—began to change my understanding of reality and of God.

To become a mechanical engineer, I learned a lot in school about the physical sciences. Like many engineers, before I would believe something, I needed proof. And proof of the spiritual world is pretty hard to come by. There is no evidence that proves the existence of the spiritual world using internationally accepted scientific methods. There is, however, a lot of evidence pointing to a creator, and that's what this book is all about.

I grew up in the church and was exposed to the writings of the Bible many times. As a curious teenager, I read the entire New Testament and learned a great deal about what the Bible teaches. Based on my understanding of the physical world at the time, the Bible seemed crazy to me:

A man swallowed by a whale, living in his belly for days, and being spit out alive? *C'mon, get real.*

Demons being cast out of people? *Sounds like ignorance of the psychological sciences.*

Jesus raising people who had been dead for days? *Exaggeration and folklore.*

God knows everything at the same time. And he even knows what will happen in the future? *How can that be possible?*

Predestination and free will? *How can both things be true when they are opposites?*

It turns out that the Bible describes a person with a belief system like I had at that time:

The person without the Spirit does not accept the things that come from the Spirit of God but considers them foolishness and cannot understand them because they are discerned only through the Spirit. (1 Corinthians 2:14)

And as crazy as these Biblical statements and stories seemed to me, I came to learn that they are no crazier than our physical universe.

> The more I learned about physics, the more I learned that our world isn't what it seems to be.

Physicists have known crazy scientific facts for most of the last century, but most physicists aren't interested in the theological and philosophical implications of them. Most don't seem very interested in spreading the word to non-scientists either, so most people in the world don't understand the craziness that takes place at the subatomic level (very small) or the astronomic level (very large).

And when you do bump into a physicist who is happy to share what she knows with you, you probably won't be able to understand her. Most physicists speak a language that "normal" people can't understand.

INTRODUCTION

In this book, I will do my best to use easy-to-understand language and concepts to share some of the information that I learned in my lay person's study of modern physics and how it might relate to faith.

It is important to point out that the facts and theories in this book come from established physicists like Albert Einstein, Richard Feynman, Stephen Hawking, and Brian Greene. These scientists are not prone to believe in fringe theories. The physicists I reference believe in things that can be consistently duplicated in controlled laboratory experiments. I did not include any fringe theories in this book, as I believe that these kinds of theories can be very dangerous to promote and explore.

The concepts I will cover don't prove anything about my faith system or yours. However, they do point to evidence that everything we believe in, whether physical, logical, or spiritual, is based at some level on faith.

What you will find in this book is the information that rocked my world and allowed me to take the leap of faith I needed to accept Jesus as my Lord and Savior.

THE LAYOUT OF THIS BOOK

This book is laid out in five chapters:

Chapter 1: What Is Faith?

Chapter 2: The Size of Things

Chapter 3: The Light of the World

Chapter 4: The Illusion of Physical Reality

Chapter 5: What It All Means

I have done my best to make this information easy to understand, but some concepts are inherently more complicated than others. Therefore, each section starts with relatively basic concepts that are easy to understand and progresses into the more complicated science, which might prove to be challenging for some.

I hope you find this deeper dive into faith and physics interesting and I pray that the Holy Spirit will use it for your good just as was done for me.

A NOTE ON LANGUAGE

I like simple language and sometimes lose a little accuracy because of it. I'm okay with that in this case because I'm trying to help lay people understand some basics of what physicists—those who deal with great precision and very complex words and mathematical equations—have learned about how the universe works.

This is also true of my theological language. I use simple theological language, which may or may not be 100 percent accurate. I will refer to God at times as "he" or "him." I don't really think of an infinite God as male or female and I am not trying to suggest either.

I hope this won't be an issue for you, because it's not pertinent to the challenging science and theology that we'll be exploring. I just prefer the simplicity of pronouns for this book so it can be as straightforward as possible for a non-academic reader.

WHAT IS FAITH?

"Faith is to believe what we do not see, and the reward of faith is to see what we believe."

—Saint Augustine

Before we get into the physics that rocked my world, I would like you to consider faith. Specifically, what is faith and how does it affect our daily living?

DIVE DEEP

Grab a pen and a piece of paper and spend some time thinking about what faith is and how it impacts our lives on a daily basis. Don't limit this to your own experience. Think about faith on a broad scale. What is it? How does it affect us?

So, what is faith? Simply put, faith is belief. If you believe in something, you have faith in it. From a Christian perspective, faith is belief in God the Father, Jesus the Son, and the Holy Spirit. The dictionary has several slightly more complicated definitions, but I want you to understand how I view faith so you will understand how I am using the term in this book.

Let me share a few of the hundreds of Bible verses that refer to faith:

Now faith is confidence in what we hope for and assurance about what we do not see. (Hebrews 11:1)

[Jesus] replied, "You of little faith, why are you so afraid?" Then he got up and rebuked the winds and the waves, and it was completely calm. (Matthew 8:26)

[Jesus speaking to Peter as he tried to walk on water] Immediately Jesus reached out his hand and caught him. "You of little faith," he said, "why did you doubt?" (Matthew 14:31)

[Jesus] replied, "Because you have so little faith. Truly I tell you, if you have faith as small as a mustard seed, you can say to this mountain, 'Move from here to there,' and it will move. Nothing will be impossible for you." (Matthew 17:20)

These verses are just a few of many in the New Testament that refer to our faith, and more specifically to our lack of faith. It is apparent to me that virtually all of us lack faith.

Why do we lack faith? I suspect that there are several reasons.

One is because of the earthly reality in which we find ourselves. Have you ever watched a scary movie and had to remind yourself that it's not real? This is a good example of how the *perception* of reality that is created by your five senses has a big impact on how you experience *true* reality. Even though you know the movie isn't real, emotionally you probably respond as if it were. You have to remind yourself continually that the movie isn't real. Imagine if that movie were 24 hours a day, seven days a week, 365 days a year, and it was three-dimensional and entirely immersive. That would be a very powerful movie!

And imagine that you were born into that movie house so your very first experiences of consciousness were bathed in this earthly reality—this movie reality. If there is a reality outside of this movie you have been immersed in from the very beginning of your life, it would be very difficult to know it and to stay in touch with it. You'd have to remind yourself that "it's just a movie" constantly, metaphorically speaking.

Paul of Tarsus says these words in Romans 12:2:

Do not conform to the pattern of this world, but be transformed by the renewing of your mind. Then you will be able to test and approve what God's will is—his good, pleasing and perfect will.

WHAT IS FAITH?

I believe that this world, which Paul says we are not to be conformed to, is not the real deal. Heavenly reality is! We are to tell ourselves continually that it's "just a movie." This is how we are transformed by the renewing of our minds.

> We need to remind ourselves constantly that there is a greater reality, a heavenly reality, which changes who we are and how we live.

But even if we choose to believe in this greater reality—in God, in Jesus, and in the Holy Spirit—we are bombarded every day with earthly things, and it is easy for our automatic responses to be earthly instead of heavenly.

We must constantly remind ourselves that the kingdom of heaven is the one true reality. This is one reason why a practice of daily focus on God is so important. It is similar to the practice we use when watching a scary movie: continuously reminding ourselves of the one true reality.

The Bible teaches us not to worry about the future, about food, about our physical well-being—all of these are in God's hands, and he has promised to provide for us, to protect us, and never to forsake us.

However, our earthly world doesn't tell us these things. We naturally focus on key survival strategies such as stockpiling resources for unforeseen crises, saving money for retirement, and protecting our resources for fear that we could go without at some point. And while these things may be good to do for our walk in this world, I don't think this is what Jesus was primarily describing when he spoke about our walk in faith.

How many of us truly trust God and believe that he will take care of us? If we did—if we truly believed that God is in control of everything that happens—then we would trust him to calm all of the storms in our lives or give us the strength and wisdom to get through these storms. After all, God promised us these things, and if God is with us, who can be against us? Most Christians, me included, struggle to have a faith that is this strong, and yet most Christians *profess* to have faith that is this strong. What causes this disconnect?

DIVE DEEPER

How strong is your faith in God's promises? What storms are raging in your life right now that you need God to calm? What are some ways you can practice being faithful?

What are some of the things that God promises us in the Bible?

- A life of abundance

- To be with us to the end

- God's peace and comfort

- Eternal life in heaven with God

- Rewards in heaven for the suffering experienced on earth

- Rewards in heaven for the unconditional love and blessings offered to those who may never recognize them

There are too many promises to list here. There are thousands of God's promises in Scripture, and the list above is just a start.

Why is having real faith so difficult for us? Why is it so hard to trust in the promises that God has made to us? Perhaps we struggle because so many faithful people experience tragedy, illness, and persecution. We wonder, "Where is God in all of this? How could God let this happen?"

We know that people of faith suffer in this world. Think about all that happened to the apostles. Did God take care of them? From an earthly standpoint he did not. They were persecuted and killed. And yet we Christians choose to believe that they were completely protected by God.

> Maybe our well-being, which God tells us is in his hands, is more spiritual than physical.

Maybe our faith is to be based on our heavenly reality more than our physical reality—a protection of our souls more than our bodies.

But ask yourself: on a day-to-day basis what seems more real to you, your body or your soul? I suspect that for most of us, our bodies seem more real. However, the answer might differ if we asked this question of someone whose body has begun to fail or who has experienced a significant spiritual event.

The point of this brief exploration of faith is to point out that for most of us, the physical world *feels* more real than the spiritual world. What a gift it would be to experience the spiritual world more fully than we do the physical world. Then we could live by sight, not just faith.

What we experience in the physical world isn't as real as we think.

Modern science tells us that much of what we experience is some form of illusion.

Interestingly enough, the Bible also tells us—in fact warns us—not to be fooled by earthly things. It speaks of earthly things as if they are illusions of ultimate truth.

The Spirit gives life; the flesh counts for nothing. The words I have spoken to you—they are full of the Spirit and life. (John 6:63)

Set your minds on things above, not on earthly things. (Colossians 3:2)

Paul tells us not to be conformed to the things of this world, but to be transformed by the renewing of our minds. My prayer is that the following pages will help you do just that.

So, let's dive into God's creation and see just a bit behind the curtain into the illusion of this physical world that scientists learned about in the twentieth century.

THE SIZE OF THINGS

From the full expanse of the universe to the smallest of particles, we just happen to be right in the middle of it all.

Some of the facts I will share with you are amazing—mind-blowingly amazing. Others are bizarre and mind-boggling. I hope that engaging in this material will help you appreciate how amazing and bizarre this physical world of ours really is.

Some of the facts of our universe can only be described with numbers and sizes that are too big for our minds to wrap around. Other parts of our world are so counterintuitive, so unlike anything we thought could be true, that our minds just can't understand them.

This section of the book is about the really big stuff and also the really small stuff in our universe. The numbers and sizes are so mind-blowing that these concepts will impact the way you see God. This chapter will help you see how incredibly small and insignificant you are and at the same time how massive and huge you are.

HOW BIG IS BIG?

Have you ever stood next to a mountain and felt so tiny that it seemed impossible that the creator of that mountain could even know you exist? I have felt this several times. But that mountain is nothing compared to our universe.

As an illustration of our relative size, let me talk about the amazing mountains in Yosemite Valley like Half Dome, El Capitan, and all the others that Ansel Adams made so famous with his beautiful photography. I've stood in Yosemite Valley two different times in my life. Each time I felt completely dwarfed by the huge ponderosa pines, which were themselves completely dwarfed by the mountains. Picture this in your mind if you can. This is one frame of reference.

Yosemite Valley, California

Now, here's another: When you look at one of the first pictures taken of the Earth from the Apollo space missions, you can't even come close to seeing Yosemite Valley. Or better yet, when you look at a picture of Earth taken from one of the Mars orbiters, not only can you not see Yosemite Valley, but from Mars the Earth looks smaller than the moon normally looks to us. Or when you look at a picture of Earth taken by the Cassini probe near Saturn, our Earth is just a tiny pinprick of a dot in the photo. How big is Yosemite from that perspective?

THE SIZE OF THINGS

View of Earth from Apollo 17 (1972)

View of Earth from Mars Orbiter (2007)

View of Earth through Saturn's Rings from the Cassini Probe (2006)

View of Saturn from Cassini (2011)

And this distance is **nothing** when looking at the size of our galaxy.

The Earth and all the other planets in our solar system orbit around the sun, which is just an ordinary star in the Milky Way galaxy. Scientists estimate that there are 400 **billion** stars in our galaxy.

That's 400 thousand million.

THE SIZE OF THINGS

Stars in our galaxy seen through the Hubble Space Telescope

Think about that.

The University of Michigan's football stadium can hold almost 115,000 people. I've been there, and I can tell you that's a lot of people! I'm told that the Indianapolis Motor Speedway, where the Indy 500 is held, seats around 400,000 people if you count the stands and the infield. So imagine that huge crowd being assembled a thousand times over. Put all those people together and you'd have 400 million people—roughly the population of Europe.

Now take that unbelievably large crowd of 400 million and multiply it a thousand times over to get 400 billion people. The current population of the Earth is around seven billion. The population of the stars in our Milky Way galaxy is **400 billion**!

It is mind-blowing to me that a being could create this kind of grandeur. Do you believe in a God who is this big and mighty?

Now let me give you a sense of not just the numbers, but the size of our galaxy.

If I could shrink our galaxy—way down until the sun became the size of a light bulb—how far away would you guess the next light bulb (star) would be (which, by the way, is called Alpha Centauri)? What do you think? One hundred yards away? A mile away?

The next star would be about 300 miles away. And in between these two stars would be nothing—just 300 miles of space between two light bulbs.

For a being to have created 400 billion stars and all this empty space in between is so mind-boggling that it was hard for me to believe there could be a creator this big and mighty.

Let's get back to Yosemite Valley. Where do I fit into all of this? I'm blown away by the mountains, which are tinier than tiny in the whole scheme of things. I'm guessing it blows your mind too.

But I'm going to give you more mind-boggling facts in order to drive home my point about how big and mighty God is.

The M16 Eagle Nebula, an enormous stellar nursery, lies 7,000 light-years away in the inner spiral arm of the Milky Way Galaxy

THE SIZE OF THINGS

Our Milky Way galaxy of 400 billion stars seems to be an average galaxy. Using the Hubble Space Telescope, scientists have been able to estimate the number of other galaxies in our universe. Can you guess how many galaxies scientists are aware of? Dozens? Hundreds? Millions?

A magnificent spiral galaxy

There are over 100 **billion** galaxies.

Two spiral galaxies

I'm not sure what to say to all of this, so I'll just share with you what King David said in Psalm 8 (verses 1, 3, 4 and 9):

LORD, our Lord, how majestic is your name in all the earth! You have set your glory in the heavens.

When I consider your heavens, the work of your fingers, the moon and the stars, which you have set in place, what is mankind that you are mindful of them, human beings that you care for them?

LORD, our Lord, how majestic is your name in all the earth!

How small do you feel right now? How big is God to you now? Don't you think we sell God a little short when we don't understand how mighty and powerful he must be to have created all of this?

Praise God! Selah! (As King David would say.)

The Large Magellanic Cloud galaxy

THE SIZE OF THINGS

Take some time to ponder how big the universe is. How does this make you see God differently? How do you think your problems look to God?

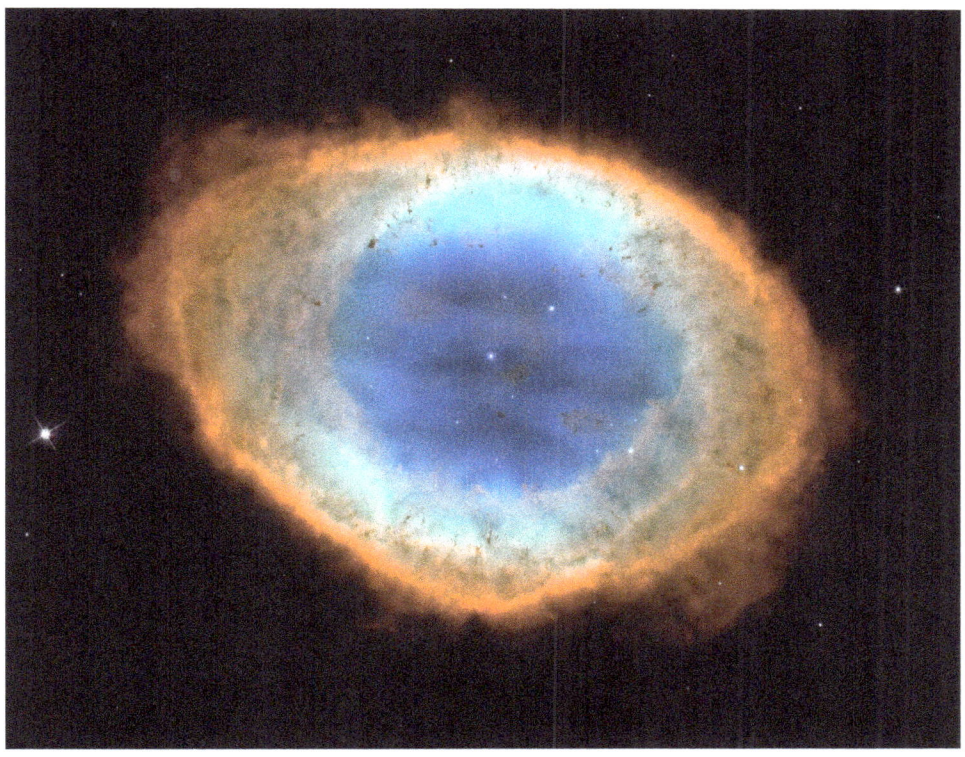

Messier 57, The Ring Nebula

The great clouds of the Carina Nebula in the Carina–Sagittarius Arm of the Milky Way

The Pillars of Creation from the M16 Eagle Nebula

THE SIZE OF THINGS

HOW SMALL IS SMALL?

As small as you feel right now compared to the grandeur of the universe, that feeling is reversed when you're compared to the smallest things scientists are studying.

One of the smallest particles scientists study is the neutrino. A neutrino is a very tiny particle—much, much smaller than the nucleus of an atom. It's so small that the ratio of the entire universe to you is the same ratio as you to the neutrino. We are right in the middle of the very, very large things and the very, very small things that we're able to observe.

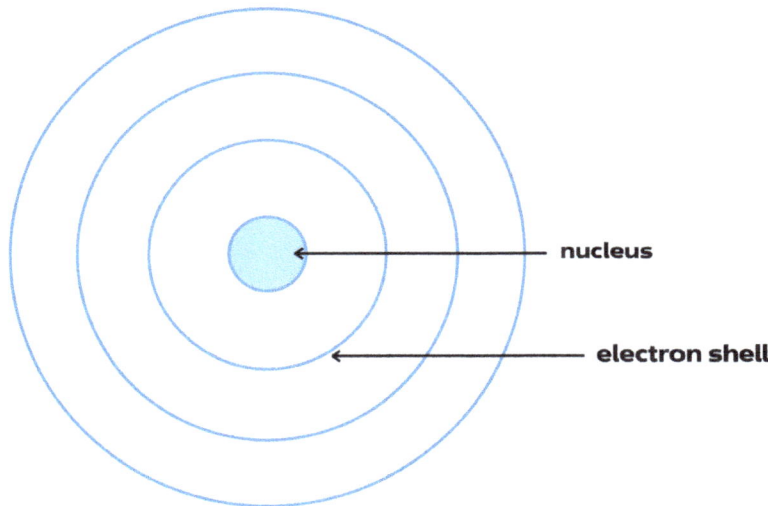

An atom is about 100 million times smaller than you. At the core of an atom is the nucleus, which is about ten thousand times smaller than the entire atom of which it is a part.

And just like the vast emptiness I described between stars and galaxies, in the gap between the nucleus—one ten-thousandth the size of the overall atom—and the outer boundaries of the atom in which the electrons reside is nothing.

Atoms are 99.999 percent nothing—at best.

The electrons of an atom, whose location determines the outer diameter of the overall atom, are ten thousand times smaller than the nucleus.

To put this into perspective, imagine a golf ball at the center of the 50-yard line of a football stadium. The golf ball represents the nucleus of an atom. The electrons hanging out at the outer edge of the atom are one-ten-thousandth the size of the golf ball—say one-tenth the size of a grain of sand in this model—and would be located outside the stadium in the parking lot.

The relative size of a hydrogen atom

In between the golf ball and the micro-grain of sand is nothing.

Nada.

In this analogy, the next closest atom would be another golf ball in another stadium maybe 150 yards away, and between these golf balls would be a few micro-grains of sand (the electrons) and nothing else.

THE SIZE OF THINGS

In inner space, there's a whole lot of nothing! And earlier we learned that in outer space, there's also a whole lot of nothing. It's only in our limited world in the middle where there appears to be a lot of something.

And it keeps getting weirder: When scientists look deeper into the nucleus of an atom to find out where the real matter is—you know, the real stuff that makes up the material world—they just keep finding nothing.

Is it possible that a being actually created the hugeness of the universe AND the smallness of each atom that makes up everything in the universe? And is it possible that this being created all this "solid" stuff around us out of a whole lot of nothing?

Atoms are 99.999% empty space

If so, this is an amazing being. This being would be far beyond what our minds could even hope to understand. As the prophet Isaiah said in Isaiah 46:5:

With whom will you compare me or count me equal? To whom will you liken me that we may be compared?

Well said, Isaiah! And that's about the size of it!

Our world, which seems so full of "stuff," is really full of a whole lot of nothing. What does this tell you about the role God plays in holding all things together? Take some time to reflect on Colossians 1:17. *"He is before all things, and in him all things hold together."*

THE LIGHT OF THE WORLD

This is the message we have heard from him and declare to you: God is light; in him there is no darkness at all. (1 John 1:5)

In him was life, and that life was the light of all mankind. The light shines in the darkness, and the darkness has not overcome it. There was a man sent from God whose name was John. He came as a witness to testify concerning that light, so that through him all might believe. He himself was not the light; he came only as a witness to the light. The true light that gives light to everyone was coming into the world. (John 1:4–9)

When Jesus spoke again to the people, he said, "I am the light of the world. Whoever follows me will never walk in darkness, but will have the light of life." (John 8:12)

To the physical scientist, light is one of the most mysterious things in our universe. It is also one of the most common metaphors used by God to describe himself. There is so much more to light than you ever imagined.

Let's review a few of the bizarre facts about the properties of light that scientists have discovered:

- The speed of light is an unchanging standard unlike anything else in the universe. It never varies, regardless of the relative speed of the person or machine that observes it.

- Time has no meaning for light. Time flows differently for objects depending on how fast they are moving. Time stands still for anything that travels at the speed of light. Therefore, if time doesn't flow for light, then light touches everything at the same time.

- Light has a distinctly dual nature; it is both a particle and a wave, which are opposites of each other.

- Light acts differently depending on how we observe it. Somehow it seems to be affected by how much we know about its behavior.

- Light seems to predict how we will observe it, and it acts accordingly ahead of time.

What follows is a review of the theories and experiments that support these crazy statements.

LIGHT: A STANDARD UNLIKE ANY OTHER

In 1905, Albert Einstein published his special theory of relativity. This theory suggests that, among other things, the speed of light is always the same in a vacuum, regardless of how fast the observer of the light might be moving.

Nothing else in the universe does this.

Unlike any other thing that moves, light's speed is not relative to the observer's perspective.

Let me explain: We tend to judge the speed of objects against a background of stationary things such as trees, buildings, or cornfields. However, if an object is moving in outer space and there are no stationary objects with which to compare its speed, it is impossible to say what the speed of the object is. (Remember how empty space is? There aren't many objects to use as references to judge speed.) The measurement of speed is a relative thing.

Consider a spaceship that appears to be moving 10,000 miles per hour toward another spaceship. Without any other objects in the picture to reference the spaceships against, you can't say whether spaceship A is moving toward spaceship B at 10,000 miles per hour, or if spaceship B is moving toward spaceship

A at 10,000 per hour, or if they are both moving toward each other at 5,000 miles per hour.

The forces we feel when we travel are due to acceleration, not our steady speed, so the spaceship travelers, who are moving at a constant rate of speed, would feel no acceleration and have no way to determine who is moving at what speed. All the travelers know for sure is that their speed relative to each other is 10,000 miles per hour.

Here's another example, one that is more down to earth. Imagine you are in a car moving at 50 miles per hour alongside another car going 80 miles per hour in the same direction. You would perceive the car next to you to be going 30 miles per hour relative to your car. However, because of the stationary objects nearby, you would be able to see that the other car—and you—are going much faster than that.

Conversely, if you were in a car going 50 miles per hour toward another car approaching at 80 miles per hour, without stationary objects around you, you would perceive yourself as standing still and the other car coming toward you at 130 miles per hour. At the same time, the driver of the other car would perceive herself as standing still and you moving toward her at 130 miles per hour. Without stationary objects around to help you gauge your speed, it's all relative.

Einstein's theory of relativity states that the relative nature of speed exists for all things that move.

Except light.

Regardless of how fast or slow an observer is moving, the speed of light always looks and measures the same: 186,300 miles per second. Even if you are moving at 180,000 miles per second alongside a light beam, if you take a moment to measure it, you will perceive light as traveling at 186,300 miles per second.

Intuition and common sense would lead us to believe that the light wave's speed would only measure 6,300 miles per second. But the speed of light always measures the same—186,300 miles per second—regardless of the speed of the observer.

What about sound waves? The speed of sound works the same as the cars in my illustration above. In normal conditions, sound travels at around 750 miles per hour. If a person traveling at 700 miles per hour came alongside a sound wave and measured the speed of that wave, it would measure at just 50 miles per hour, because the speed of sound is relative.

As you know, some jets can travel faster than the speed of sound, which means that the sound created by the jet can't keep up with the jet itself. This is the moment when a sonic boom occurs.

The speed of light is much, much faster than the speed of sound: Sound can travel 750 miles in one hour. In that same hour, light will travel 670,680,000 miles (186,300 miles per second x 60 seconds in a minute x 60 minutes in an hour).

That's a huge difference.

This is why sight and sound can become out of sync when you observe things that make noise from far away. Examples are lightning and thunder, firework booms and blooms, and a baseball bat hitting a ball when observed from deep in center field. In each of these cases the light that you see gets to your eye lickety-split, while the sound moseys at a mere 750 miles per hour.

The fact that light's speed is very, very fast isn't anything radical, but the fact that it always measures the same, regardless of the observer's speed, is radical.

Beyond radical.

It's amazing to me that Einstein would even think up a crazy theory like this, but he did. Along with it are some amazing consequences that we will review. But before I move on to cover these consequences, I want to restate this amazing fact:

There is only one thing in the physical universe that behaves this way. Only one thing is an unchanging standard regardless of our perspective.

THE LIGHT OF THE WORLD

And that one thing is light.

There is an interesting parallel between light and God. I love the fact that light is so often used as a metaphor for God, for he is also a standard unlike any other. It doesn't matter *when* you experience him, *where* you experience him, or *how* you experience him. God is always the same.

We change, and sometimes we feel that God has changed, but when we approach him, he is always there to love us and give us hope, peace, and joy. Though our world has changed immensely over hundreds and thousands of years, God is still the same. Just like light, he isn't a relative standard; he is an absolute standard!

While I am in the world, I am the light of the world. (John 9:5)

This is the message we have heard from him and declare to you: God is light; in him there is no darkness at all. (1 John 1:5)

The sun will no more be your light by day, nor will the brightness of the moon shine on you, for the LORD will be your everlasting light, and your God will be your glory. (Isaiah 60:19)

The Bible says that God never changes and will always love us. What are some of the unchanging characteristics of God that give you comfort?

TIME FLOWS DIFFERENTLY FOR LIGHT

Einstein's theory has been tested in several different ways, and it has always passed with flying colors (no pun intended . . . well, maybe just a little). Upon deeper investigation of this theory, there are a couple of significant implications:

1. If you travel at a significant percentage of the speed of light, time slows down for you as compared to those not traveling that fast.

2. If something could travel as fast as the speed of light (which Einstein believed was impossible for anything that has mass), time would stop altogether for that object.

I suspect that if you had never heard this before, it might be a little unsettling. When I share this fact with friends, I'm often asked, "You mean the rate that time flows isn't constant for everyone and everything?"

No, time is not constant. The rate that time flows at depends on how fast you are traveling through space.

THE LIGHT OF THE WORLD

In the 1970s, scientists conducted an experiment with one of the fastest jets available and two of the most accurate clocks ever built. These incredibly accurate atomic clocks were synchronized to each other, and then one was placed on the jet while the other stayed on the ground. The jet then flew around for a few hours at top speed. After the completion of the trip, the two clocks were compared to see if they were different.

The clock traveling on the jet recorded a little less time than the clock that stayed on the ground, which means that time slowed down for the clock traveling in the jet. The difference between the two clocks was exactly what the math behind the theory of relativity predicted it would be.

But do not forget this one thing, dear friends: With the Lord a day is like a thousand years, and a thousand years are like a day. (2 Peter 3:8)

After over 100 years of examination and experimentation, today physicists treat this radical theory of the constant speed of light as fact.

LIGHT TOUCHES EVERYTHING AT ONCE

Now let's explore what the Theory of Relativity has to say about traveling as fast as the speed of light.

Though Einstein surmised that nothing that has mass could travel at the speed of light, there is one massless entity that can. By an amazing coincidence, light travels at the speed of light (sorry, this is my attempt at Yogi Berra-ish humor). Therefore, as the theory states, time must stand still for light.

To put it another way, time has no meaning for light.

Once again, this might be a little unsettling for you. Welcome to modern physics!

Let's look into this light-time thing a bit further.

- If light travels from point A to point B and no time elapses for light, then from the photon's perspective, the light touches point A and point B at the same time (a photon is the smallest packet of light that can exist).

- Light comes to us from other galaxies, which takes thousands, millions, or billions of years to reach us. However, to that photon of light not even one second ticked off from its internal clock. Therefore, everything that the photon of light touches, it touches at the same time.

Getting back to the original question that Einstein asked about light, how does the Universe "look" to a beam of light (or a photon, if you prefer), or to a person riding on a light beam? And how does time flow for a photon?

You can either say that time does not exist for light, which means that light is everywhere along its path (everywhere in the Universe) at once, or you can say distance does not exist for light, which means that it touches everything in the Universe at once.

The following is an excerpt from *Schrödinger's Kittens and the Search for Reality* by John Gribbin, the sequel to *In Search of Schrödinger's Cat*:

THE LIGHT OF THE WORLD

Schrödinger's Kittens and the Search for Reality

By John Gribbin

"From the point of view of the photon, of course, it is everything else that is rushing past at the speed of light. And under such extreme conditions, the Lorentz-Fitzgerald contraction reduces the distances between all objects to zero. You can either say that time does not exist for an electromagnetic wave, so that it is everywhere along its path (everywhere in the Universe) at once; or you can say that distance does not exist for an electromagnetic wave, so that it 'touches' everything in the Universe at once." (Page 79)

Knowing this, we could describe light as "omnipresent."

Sounds like God! He is everywhere at once. Time doesn't apply to him as it does to us. He is the alpha and the omega.

Think about how many of our problems and concerns are related to time. Think of a recent problem you have experienced and how time played a role in it.

Whether suffering from illness, a bad job, or the loss of a loved one, in most cases time seems to fix our problems. Of course, timelessness would not alleviate us from having problems, but it seems that they would become more manageable.

THE LIGHT OF THE WORLD

If you could experience everything in your life at once, you probably wouldn't be so concerned about the future. What if living life could be like watching a play that you had seen before? Even though a scene would be no different than the first time you saw it, knowing what happens next would help you not get too anxious.

Just imagine how much stress would be lifted from you if only you knew how things would work out in your life. The unknown often causes much more stress than the problems themselves cause. This is why I sometimes ask, "what's the worst thing that could happen?" I've found that often the answer to that question reveals that we're less afraid of that worst thing than we are of the unknown. "A problem named is a problem tamed."

So, how might all of this relate to faith? Knowing that light is able to do things that I once thought impossible, and that are similar to things I'm told God does, makes me more confident that God is real and present in our lives! He can do all these things and more.

Does believing that things *will* work out in the future help as much as knowing *how* things will work out?

The answer can be yes.

Of course we will never be able to understand why things work out the way they do, but we are told that God knows all, that he is in control, and that in the end all things will work out for the good of those who love him.

And we know that in all things God works for the good of those who love him, who have been called according to his purpose. (Romans 8:28)

I believe that it would be easier for us to have strong faith if we didn't live within this earthly reality where time seems so real and we can't see forward to know how things will work out.

Are you waiting on God for anything right now? How does the verse above from 2 Peter make you feel about waiting on God? How might you learn to trust in God's perfect timing?

AMAZING OPPOSITES: PARTICLES AND WAVES

For over 100 years, there has been a debate about what light really is.

Is it a particle? Is it a wave?

I remember in my high school physics class, Mr. Reed and Mr. Deslitch dressed up as Wave Man and Particle Man and debated the issue in front of us. These guys loved physics and they were often weird—truly a good fit for teaching high school science! To this day I am grateful for their passion for teaching physics.

I very clearly remember their debate about whether light is a wave or a particle, but I don't remember them ever telling us the answer. And that's because to this day, scientists don't really understand what light is.

Let's review some of the thinking and experiments that have taken place around the nature of light: If light were a particle, it would behave in a similar way to a little BB bullet. If light were a wave, it would behave in a similar way to waves that you see on a lake.

There have been several ingenious experiments designed to prove one way or another if light is a wave or a particle. And what scientists have found is that light acts like a wave when the experiment is designed to show that it is a wave. But it acts like a particle when the experiment is designed to show that it is a particle.

What?

It's no wonder that scientists disagree about the nature of light. Scientists may not understand what light is, but they do understand how it behaves. And the way it behaves is pretty spooky. I'll get to this spooky behavior in a bit, but first I need to describe one of the key characteristics of waves called *interference*.

WAVE BEHAVIOR AND INTERFERENCE

To understand how waves move, think about the waves you see on a lake. If you were to drop a stone into a smooth pond, you could watch the ripples—waves—move toward the shore. You see the motion of the wave. The particles of water, however, only move up and down. They don't move toward the shore, or at least not much.

One of the unique things about waves is that when two of them come together, they interfere with each other. When peaks combine, they add together, so two one-inch wave peaks combine to become a single two-inch wave peak. The same is true of two troughs that combine: Two one-inch wave troughs combine to form a single two-inch wave trough. This is called "constructive interference." When a one-inch peak combines with a one-inch trough, they cancel each other out as if there was never a wave in the first place. This is called "destructive interference." The figure below illustrates both destructive and constructive interference.

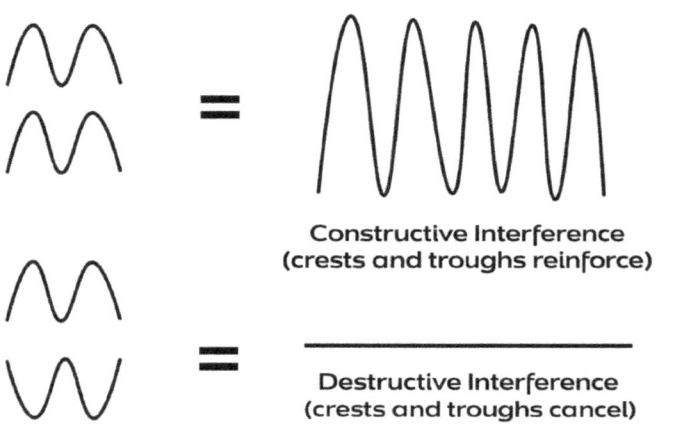

You could see evidence of interference on the side of a pool if you threw two stones into two different areas of the pool simultaneously. Once the water calmed down, you could look at the wet marks the water left on the sides of the pool. There would be areas where the wet marks indicate waves that were twice as high as expected due to constructive interference, and other areas without any wet marks because destructive interference occurred and the waves canceled each other out.

Another common experience of wave behavior occurs with radio signals in large cities. Have you ever been listening to a clear radio station that suddenly becomes staticky when you stop at a traffic light? If you move just a foot forward, the reception will become clear again. This is a result of the radio waves bouncing off buildings and experiencing constructive and destructive interference. For our purposes it's not important that you understand exactly how wave interference works, just that you understand that wave interference exists.

Now let's move to the experiment that is at the center of much of the weirdness with modern science. To this day, this experiment baffles everyone who tries to understand it.

THE DOUBLE-SLIT EXPERIMENT

To determine if light is a wave or a particle, scientists devised the apparatus shown in the following figure:

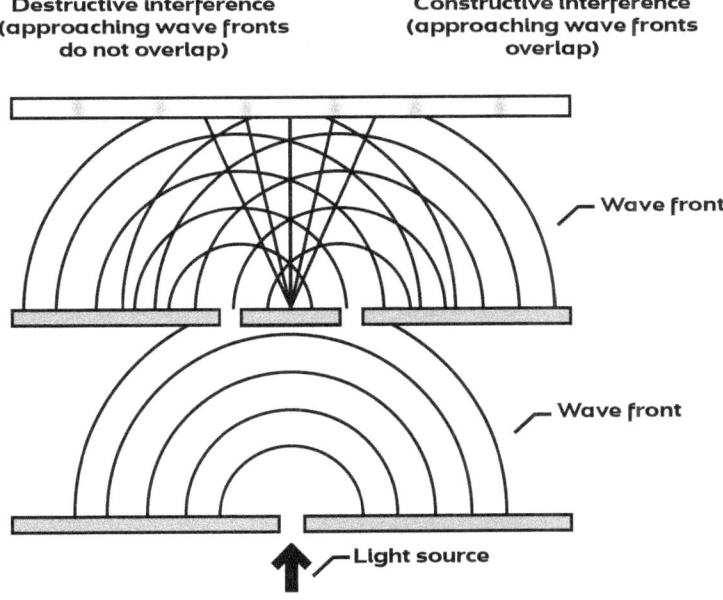

If light is a bunch of small particles, or BBs, each particle should go through one of the two slits in the middle wall in a straight line and not interfere with itself before hitting the back detection wall. However, if light is a wave, then two waves will result from the two slits, and these two waves will interfere with each other, just like the waves in the swimming pool analogy I shared with you earlier. This interference would be detected on the back wall of the apparatus, just like interfering waves can be detected on the side wall of a pool (as illustrated in the figure above).

The initial experiment was run and evidence of interference was found. The scientists who devised this experiment determined that light must be a wave, because if light were made up of particles, there would have been no interference pattern detected, just one neat line behind each of the two slits where the photons had hit the back detection wall.

However, because other scientists had seen experiments where light behaved like a particle, they did some additional versions of the double-slit experiment.

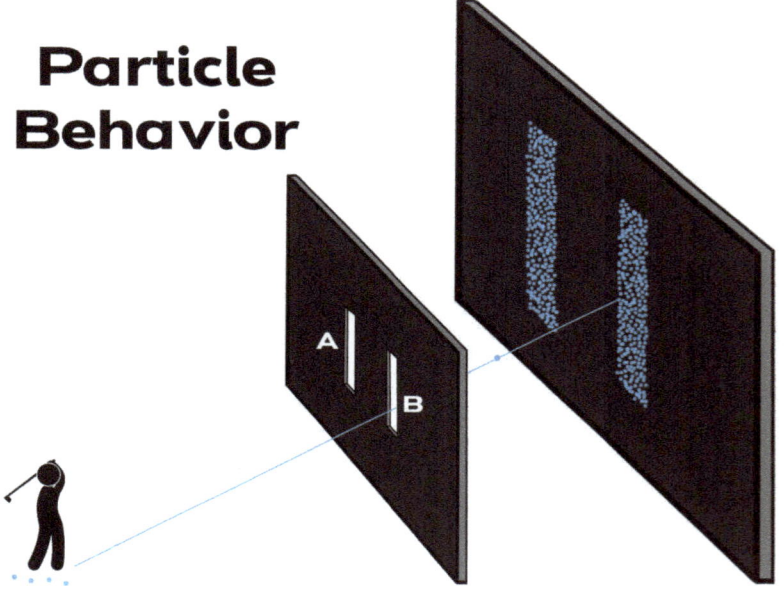

Scientists had known for some time that light could be divided into discrete packets called photons. Photons of light cannot be divided any further. A half of a photon is an impossible thing. In this way, light is like a particle, not a wave. The scientists who saw particle behavior in previous experiments decided that light must behave like a wave when there are many photons present, but they believed that when dealing with a single photon, it must behave like a particle.

So, a group of scientists decided to see what would happen in the double-slit experiment if only a single photon at a time was allowed to pass through the apparatus, and then another, and another, always one at a time, until thousands of photons had completed the journey and left their marks on the back detection wall. Surely, each individual photon would not be able to go through both slits at the same time and therefore could not create two interfering waves as was seen when many photons were present at the same time. It was only logical to deduce that each photon of light would go through one slit in a straight line, hit the back wall, and not create any interference.

The experiment was performed, and scientists were surprised to find that there was still evidence of interference, which could only result if the individual light photons were behaving like waves, not particles. This seemed impossible to the light-is-a-particle group and was certainly a blow to their theory.

Still perplexed, scientists wanted to try one more thing. Because they couldn't understand how one photon could go through both slits in the double-slit experiment (very much like a wave) and still be one photon (very much like a particle), they decided to watch every photon to see which slit each one went through.

Here is what they found: The first photon went through the left slit, the second went through the right slit, then left, another left, then right, and so on until all the photons had completed their appointed rounds. More importantly though, they never saw a photon split in half and go through both slits. And the back detection wall showed no evidence of interference. Instead, it showed a pattern of results one would expect when doing this experiment with particles. Once again, they were perplexed. When they conducted the experiment in this way (watching to see which slit each photon went through) they found **no evidence** of photons being a wave.

What?

In the previous experiment when the photons were not observed, there was evidence of interference. It was as if the photon split in half and went through both slits at the same time.

But when scientists observed each photon as it passed through the slit, there was no evidence of interference.

But it was the exact same experiment!

The only difference was that in the first experiment the scientists weren't watching the photons, and in the second experiment they were.

This experiment has been repeated numerous times and each time the result is the same: If scientists watch each photon to observe the path it takes through the apparatus, it acts like a particle and there is no interference. If scientists don't watch the individual photons, then each photon acts like a wave that goes through both slits at the same time and interferes with itself.

It's as if the photons know we are watching!

"Anyone who has not been shocked by quantum mechanics has not understood it." (Niels Bohr)

Some scientists have described a photon as a kind of "ghost wave" that collapses into reality when we look at it. There are two very interesting points that come from this:

1. Somehow photons seem to know if they are being watched, and they act differently depending on whether or not they have an observer. It's weird to talk about photons of light "knowing" things as if they were alive, but that is the simplest way to describe photon behavior.

2. Our observation of what is going on affects the outcome of the experiment. Therefore, our observation of reality affects reality.

THE LIGHT OF THE WORLD

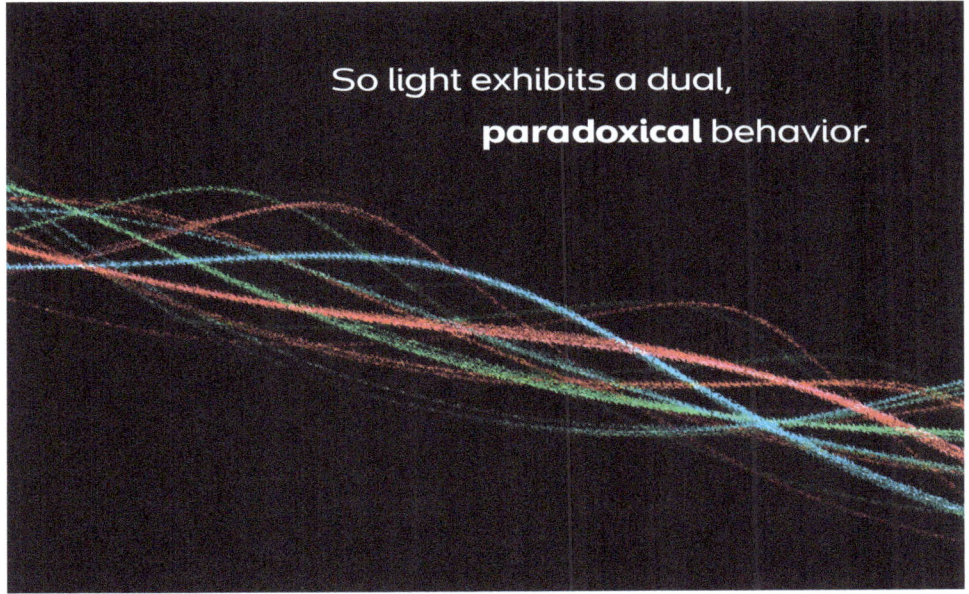

The bottom line here is that light really can be two different things at the same time: a particle and a wave. This duality seems impossible, but it is true.

Aren't there similar dualities in the Bible? We are saved by faith, but also by works.

Faith: *For it is by grace you have been saved, through faith—and this not from yourselves, it is the gift of God—not by works, so that no one can boast.* (Ephesians 2:8-9)

Works: *You see that a person is considered righteous by what they do and not by faith alone.* (James 2:24)

- Our days are predetermined, and yet we have free will.

Predetermination: *But we ought always to thank God for you, brothers and sisters loved by the Lord, because God chose you as firstfruits to be saved through the sanctifying work of the Spirit and through belief in the truth.* (2 Thessalonians 2:13)

Free will: *Now fear the LORD and serve him with all faithfulness. Throw away the gods your ancestors worshiped beyond the Euphrates River and in*

Egypt, and serve the LORD. But if serving the LORD seems undesirable to you, then choose for yourselves this day whom you will serve. (Joshua 24:14-15a)

- Jesus is fully God and fully human at the same time.

Fully human: *But when the set time had fully come, God sent his Son, born of a woman, born under the law.* (Galatians 4:4)

Fully God: *I and the Father are one.* (John 10:30)

- God is Father, Son, and Holy Spirit.

In the beginning was the Word, and the Word was with God, and the Word was God. (John 1:1)

Many of these biblical dualities are barriers to people who are trying to make sense out of the Bible. These dualities also seem to be barriers for certain denominations of faith, as these dualities have often been responsible for two camps of thinking within denominations and eventual splits. Here are a few examples:

- Denominations that believe in predestination and those that believe in free will.

- Denominations that believe works are necessary for salvation and those that believe in faith alone.

How could these opposite doctrines both be believed as true at the same time?

Could it be that the church might be unified if Christians simply understood the concept of duality—the ability to be two different things at the same time?

Maybe I'm crazy, but I think there is some wisdom buried in here somewhere.

Think about some of the dualities outlined above—faith and works, predestination and free will, fully human and fully God. Where do you stand on these issues? How has your thinking changed in learning about the amazing dual, paradoxical nature of light?

Is this all a little weird? You bet; and there's more to come.

THE LIGHT OF THE WORLD

MORE FUN WITH THE DOUBLE-SLIT EXPERIMENT

Scientists were perplexed, and still are, with the results of the double-slit experiment. Not being the type of people to sit idly when perplexed, they started devising ways to find out which slit the photon went through without really watching it—basically watching the photon without the photon "knowing" it. Maybe the machinery used to watch the photon somehow disturbed it and caused it to collapse into the BB-like particle that can only go through one of the two slits.

Of the several ingenious ways that they devised, the simplest to describe involves polarized lenses.

Before I share how polarized lenses were used in the double-slit experiment, I want to give you a brief explanation of how they work.

Light has a polarity to it. Polarity is a little tricky to explain, but the best way to envision it is to imagine light waves that go up and down, or side to side, or any combination of up and down and side to side (see the figure below for an example of an up and down light wave).

Polarized lenses are like picket fences that only allow light waves that are aligned with the fence to pass through. If the lens is polarized to align side to side, then up-and-down polarized light can't go through it. It gets blocked by the pickets of the fence.

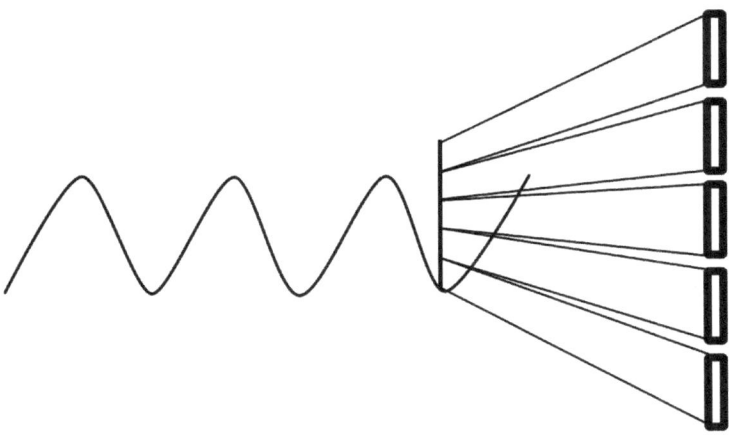

Now that you have some understanding of how polarized lenses work, let's review what the scientists did to determine which slit a photon of light went through in the double-slit experiment.

They put a side-to-side polarized lens on the left slit, and an up-and-down polarized lens on the right slit, and they ran the experiment again. This time, however, the back wall had a detection method that could determine not only where the light hit, but also what its polarity was. So, if a photon hit the back wall and it had side-to-side polarity, the scientists knew that it must have gone through the left slit because side-to-side photons couldn't pass through the polarized lens covering the right slit. This was a very clever way to record which slit the photon went through after the fact, in effect, observing without letting the photon "know" it was going to be recorded when it hit the back wall.

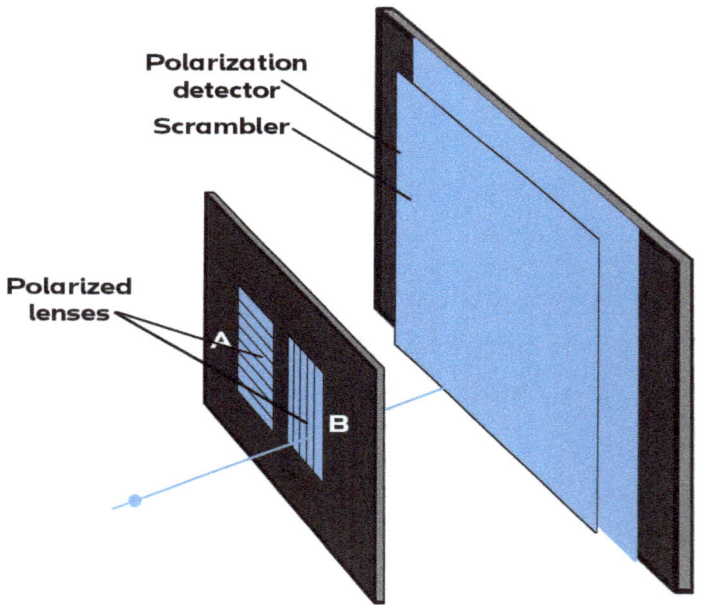

The scientists ran the experiment one photon at a time, just as they did before, but this time they could know which slit each photon went through without watching it.

What do you think happened? Was there interference or not?

Take a guess.

There was no evidence of interference. Once again, the light acted like particles, not waves. It "knew" it was being observed during the experiment.

Scientists thought that maybe the lenses caused the "ghost waves" of the photons to collapse into BB-like packets, just as they suspected happened when they observed the photons going through the slits. However, there was an uneasy suspicion that it was simply the fact of knowing that caused the light to collapse into particle behavior.

In other words, the scientists supposed that if we can have certainty of the photon's path, regardless of how we gain that certainty, the photon must act accordingly and "be" a particle, going through just one slit in the double-slit experiment and showing no signs of wave interference. If we can't have certainty of the path, then the photon must "be" a wave and go through both slits and interfere with itself.

The scientists devised one more clever variation on the polarization theme to determine if the collapse of the ghost wave was due to some interaction with the lens or if it had to do with our certainty of the photon's path.

This time they ran the polarized lens experiment exactly as before, but just in front of the back detection wall they placed a polarization "eraser." What this simple lens did was scramble the light coming through it so the scientists couldn't tell what the polarization of the light was when it first went through the slit. The light going into the lenses at the slits still came through with polarization and may have collapsed because of the interaction with the lens, but the information about which slit each photon went through was no longer available to the scientists because it had been shuffled like a deck of cards.

So, if the collapse of the ghost wave and the resulting particle behavior was caused by the lens-photon interaction, there would be no evidence of wave interference. But if the ghost-wave collapse was caused by the scientist's certainty of which path the light took, which was now erased, then there would be evidence of wave interference in the detection pattern.

Care to guess what happened?

The interference was back!

When we can know which slit it went through, it behaves like a particle.

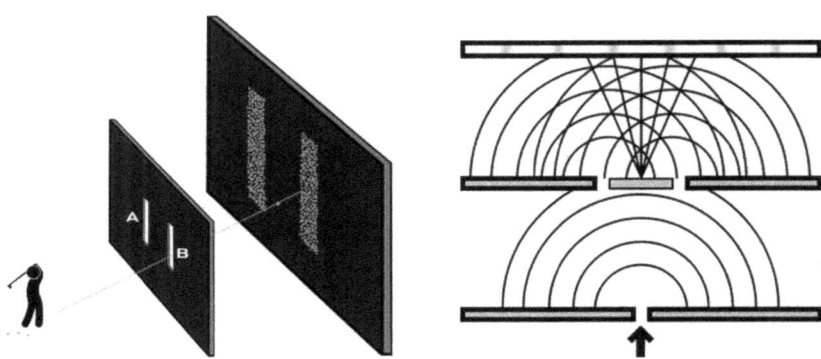

This has some startling implications:

- Light seems to know if we have information about its travels. Wow! How is that possible? Is it alive? Wow again!

- Because the actual interference takes place *prior* to the light going into the polarization eraser, somehow the light "knows" whether there is an eraser in its future or not, thus "deciding" whether to collapse and "be" a particle, or not collapse and "be" a wave. (Of course, as you learned earlier, time seems to be nonexistent for light, meaning the light goes through the polarized lens and the polarization eraser at the same time, as far as the light is concerned.)

Is your mind blown yet?

They did what your power and will had decided beforehand should happen. (Acts 4:28)

Before I took the leap of faith to believe in God, I had placed my faith in the world's beliefs. But now I was learning that my rational, logical, physical world was as crazy as what is described in the Bible:

- Light "knows" if scientists are watching in experiments and reacts accordingly (sounds like an intelligent entity).

- Light "knows" what will happen to it in the future (run into a polarization eraser or not) and it acts accordingly beforehand.

Whatever is has already been, and what will be has been before; and God will call the past to account. (Ecclesiastes 3:15)

It was at about this time in my life that I started being haunted by some biblical concepts I had been introduced to earlier in my life. Remember, I was not a believer at this time. I remembered reading that God knows the future as well as the present and the past. I had a hard time believing in a being that powerful, but then I learned that light also seems to know the future as well as the present.

The phrase "I am the light" continued to ring in my head. As I processed this new information, I couldn't help but "see the light" (once again, pun intended). The analogy between what I was learning about light and what I had heard about God was too strong for me to ignore any more.

"For you were once darkness, but now you are light in the Lord. Live as children of light." (Ephesians 5:8)

Now, I'm not saying that God is literally light. But light is used metaphorically many times in the Bible and in our Christian literature. I can almost guarantee that now you will start to notice how often we refer to God in terms like "the one true light" and "the perfect light" in our hymns, prayers, and poems.

In summary, light . . .

- is a standard unlike any other.
- doesn't seem to experience time.
- seems to be everywhere at once (a version of omnipresence).
- is both a particle and a wave, thus showing a dual nature.
- seems to be aware of what we're up to (a version of omniscience?).
- seems to know the future and acts accordingly in the present (again, omniscient?).

Characteristics of Light

1. Light appears to travel the same speed regardless of the speed of the observer (only light does this). This is Einstein's Theory of Relativity

2. Time has no meaning to light (or light touches everything at the same time).

3. Light exhibits a dual nature, (particle or wave).

4. Light behaves differently depending on how we choose to observe it.

5. Light seems to be able to predict how we will choose to observe it.

Each one of these characteristics flies in the face of common sense. They also seem to parallel what the Bible teaches us about God—the never-changing, omnipresent, omniscient, Father, Son, and Holy Ghost.

Knowing what you do now about the bizarre nature of light, how does this affect your sense of God? What questions or insights about God do you carry away from this section on light?

How does this impact your faith?

THE ILLUSION OF PHYSICAL REALITY

Now the earth was formless and empty, darkness was over the surface of the deep, and the Spirit of God was hovering over the waters. (Genesis 1:2)

By faith we understand that the universe was formed at God's command, so that what is seen was not made out of what was visible. (Hebrews 11:3)

The Son is the image of the invisible God, then firstborn over all creation. (Colossians 1:15)

He is before all things, and in him all things hold together. (Colossians 1:17)

So we fix our eyes not on what is seen, but on what is unseen, since what is seen is temporary, but what is unseen is eternal. (2 Corinthians 4:18)

In this section I will share more surprises about our physical world that have been uncovered in the past two centuries. These aren't surprises that relate to something as mysterious as light, but to things that seem more normal—everyday things like my kitchen table.

If I were to tell you that my kitchen table is solid, with no gaps or holes in it, you would quickly understand what I am describing. We all seem to experience solid objects the same way. However, as was previously shared in the chapter "The Size of Things," scientists have learned that solid objects aren't quite what they appear to be.

NOTHING SEEMS TO BE EVERYWHERE

When you look at what the universe is made of, you quickly find that it is almost all empty space—nothingness. While there are trillions of massive stars and planets, they consume very little space in the universe. Most of the universe is empty space.

In "The Size of Things" chapter, I shared the example of shrinking our galaxy down until the sun was the size of a light bulb and the next closest star was 300 miles away.

Now imagine looking down from a high-flying jet at two light bulbs that are 300 miles apart. They wouldn't be visible to the naked eye. And if everything between them were empty space, I think it is fair to assume that you would think you were looking at empty space. Except for the fact that stars are much brighter than light bulbs, and therefore much easier to see, this is what our universe is like.

Outer space is not a familiar place to us, so this fact about its emptiness might not mean much to you. But what about things that are a little closer to home, like my kitchen table? The wood in my table is made up of molecules, and those molecules are made up of atoms. In fact, all physical things, whether solid, liquid, or gas, are made up of atoms, a basic building block of our world.

Atoms are made up of two basic components: a nucleus in the center of the atom and electrons that exist in a pattern that resembles orbits. The sketch below isn't really an accurate depiction of an atom, but it's close enough for our initial exploration of it.

Obviously, the distances between the atoms that make up my table are very small. In fact, these distances can only be seen if observed through a super high-tech electron microscope. However, if we were to bring the size of a typical atom into a scale that we can relate to, similar to the "sun is a light bulb" example, you might be surprised to learn how much of an atom is nothing but empty space.

THE ILLUSION OF PHYSICAL REALITY

If the nucleus of a typical atom were the size of a golf ball, then the electrons orbiting the nucleus would be about the size of a pinprick, just barely visible to the naked eye.

No big shock here, but the location of the closest electron might surprise you.

If the golf ball–sized nucleus was at the center of one of the largest football stadiums, which holds over 100,000 people, then the closest electron would be outside the stadium in the parking lot. And in between these would be absolutely nothing.

Imagine looking at this scene from high above, the same as in the outer space example. You'd see a large stadium, a parking lot, and another large stadium. There would be a golf ball–sized nucleus at the 50-yard line of each stadium, and an electron about the size of one-tenth of a grain of sand in the parking lot of the stadiums.

Now strip away everything from the picture except the golf ball and micro grain of sand. Between those is just a whole lot of nothing. This is the nature of all atoms.

No wonder teachers don't draw atoms accurately on the chalkboard in physics class. They'd have to have a chalkboard that's hundreds of yards long.

Imagine looking at two neighboring atoms that make up my kitchen table in this scale. If you stood right in the middle of these two atoms, you would barely see a golf ball–sized nucleus about 150 yards to your right and another about 150 yards to your left. Right where you are standing you would periodically see a pinprick-sized micro grain of sand—an electron.

There is nothing else between those two atoms, and nothing else between the nucleus and the electrons of each atom.

Pure emptiness.

If you are like me when I first learned this, you are wondering what keeps solid objects, like dishes or cups, from falling right through the table to the floor, and through the floor to the basement, and through the basement to the earth, and so on.

Force fields do.

All electrons are negatively charged, and like charges repel each other, so the electrons of one atom will naturally repel the electrons of another atom.

More importantly, though, electrons only exist in tightly defined orbits, or shells, in an atom (they don't really orbit, but that works as an analogy).

These shells are rigid. They can't be compressed unless they are put under the kinds of huge forces seen in nuclear power. So, the electrons in one atom of

THE ILLUSION OF PHYSICAL REALITY

my table exist in a shell that refuses to be compressed, which is right next to another atom in my table with an outer shell where electrons exist and refuse to be compressed.

The same is true of the cups and dishes and even me. Matter appears to be solid because these electron shells cannot be compressed under normal forces. So what seems solid to us is really made up of just tightly defined force fields.

But what would we find if we dove even deeper, if we dove inside the golf ball? Inside the nucleus? It would be full of matter, right? Full of solid stuff, right?

No.

Inside the nucleus of the atom, we would once again find mostly empty space. In fact, some scientists believe that there is no "matter" at all, just highly concentrated, point-like energy fields that define everything physical in the universe. (Not all scientists believe in this point-like energy theory, so take this thought with a molecule of salt.) However, all scientists do believe that our physical world is overwhelmingly made up of empty space, both in outer space and "inner space."

> Our world, which seems so full of stuff, is made up of mostly nothing!

Apparently, my kitchen table and how I experience it is mostly determined by energy and force fields, not by the solid material that I perceive it to be. Also, when two cars collide, "solid matter" doesn't touch other "solid matter" at all. Very strong and very tiny force fields run into each other causing their geometric relationship to each other to change.

So, where does this leave us relative to the building blocks of our very existence? If everything physical is really just made up of point-like energy fields that only seem to exist when a ghost-like wave collapses into reality because of our observation of it, then what does this say about our physical world? To me it says that there is no "physical" world, at least not the way we think of it. There is an illusion of a physical world, but if you look behind the curtain you quickly find that the physical world is not what it seems to be.

By faith we understand that the universe was formed at God's command, so that what is seen was not made out of what was visible. (Hebrews 11:3)

Knowing that we perceive the world very differently from how it really is, I ask the following question: Is energy a physical thing?

As we explored in the previous chapter on light, waves are a difficult thing to wrap our heads around. When a wave created by a boat's wake travels toward the shore, what is actually moving toward the shore? The water molecules themselves just move up and down, so what do we see moving?

And what do we watch when we follow The Wave made by sports fans in a stadium? Or the perception of motion on a movie marquee or even on the movie screen? What's really moving?

When we watch atoms move up and down in such a way that we perceive something to be moving in another direction (like forward or backward), even though no "thing" is really moving in that direction, what is the reality of this? What are we seeing move?

Is the wave we're watching a physical thing?

Energy waves are certainly some kind of entity. These waves are synchronized and structured bursts of energy, but there is no physical matter in the wave, and therefore waves are not physical things, at least "not in my book" (pun intended).

Do you find this all a little unsettling? Does it make you wonder what else about our physical world might not be what it appears to be? Does this make you think of passages from the Bible that refer to things like mountains jumping into the sea?

If all of the atoms that make up a mountain are really just points of energy, why couldn't they all jump into the sea? And why don't they?

The Bible tells us that God is in control of everything. He gives order to the universe. I happen to believe this is true and that at any time God wishes, he can make a mountain jump into the sea. As you will see later in this book, these types of miracles are very possible at the atomic or quantum level of science. This means that they are also possible at our level, just very, very improbable.

I'll share more about this later in this chapter. Also, in the closing chapter I will share how I believe God, the controller of every atom, the God who knows each hair on your head, the Great I Am, uses normal subatomic science to create any abnormal event—miracle—he wishes.

But for now, let's take a deeper dive into the wonderful world of electrons.

ELECTRONS AND THE DOUBLE-SLIT EXPERIMENT

Electrons are one of the basic components of all things physical. The following things are determined by the behavior of electrons:

- Anything electrical

- All of chemistry, which means every substance known to humans, including humans themselves

- Photons of light, which are initiated and absorbed by electrons

Because electrons determine so much of our physical world, scientists thought it might be interesting to try them in the double-slit experiment instead of using photons of light.

The weird results that photons produced in the double-slit experiment left scientists reeling. But at least they had gone into it knowing that light is a mysterious entity. Electrons, however, are different. Even though they are very tiny, they do have mass and are made of matter—whatever that is. Light isn't made of matter. Scientists were quite confident that the electrons would behave like particles in the double-slit experiment, but maybe they were having a slow science day, so they decided to try it anyway.

(You might do a YouTube search on Dr. Quantum and the double-slit experiment.)

The electrons acted exactly the same as the photons. When scientists couldn't know what path it took, each electron acted like a ghost wave and seemed to exist in two places at the same time—slit A and slit B. Then each electron interfered with itself as it traveled to the back wall, just like you would expect a wave to do. When the scientists did know what path each electron took, each behaved like a normal particle that can only be in one place at one time and went through only slit A or slit B, not both.

WOW!

Our physical world is made up of electrons—physical things—that act differently depending on whether we are watching them or not.

This brings to mind all kinds of philosophical questions that I always thought were crazy. Questions such as "If a tree falls in the woods, and there is no one there to see it, did it really fall?"

You might be wondering if electrons really lose their form and become ghostly when we aren't there to see them. In the double-slit experiment they do!

THE ILLUSION OF PHYSICAL REALITY

Scientists don't know why this kind of behavior isn't noticed in the macro world—our normal sized world—but it has been repeated over and over in the micro world.

Albert Einstein, bothered by all of this, was quoted as saying, "I like to think that the moon is there even if I am not looking at it."

Now the earth was formless and empty, darkness was over the surface of the deep, and the Spirit of God was hovering over the waters. (Genesis 1:2)

When I first learned about the strange behavior exhibited by photons and electrons in the double-slit experiment, I hoped that I would later find out that scientists had made a mistake, or didn't understand some key concept, or something—anything—that would let everything go back to "normal." But alas, the more I read about how much they had studied and repeated and examined these discoveries, the more obvious it became that modern scientists are convinced this is the reality of the world.

The world in which I had placed so much of my faith was crumbling before my eyes. As I described, I struggled to take the leap of faith and believe that Jesus is who the Bible says he is because, mainly, that idea didn't match up with the world I had experienced. And suddenly I was learning that the world I knew was a fraud. The world I experience every day didn't match up with the very building blocks it rests on. The discoveries of the double-slit experiment seemed better matched to the description of reality offered in the Bible than that offered from the classical physics described by Galileo and Newton.

I am the LORD, the Maker of all things, who stretches out the heavens, who spreads out the earth by myself, who foils the signs of false prophets and makes fools of diviners, who overthrows the learning of the wise and turns it into nonsense. (Isaiah 44:24b-25)

Most scientists don't use faith in God as a way to understand these weird findings. They look for other theories that will explain their findings in some "normal," nonreligious way. Unfortunately, the theories they are playing with, theories like the existence of an infinite number of parallel universes and the existence of time travel, tend to be more bizarre than what is described in the Bible.

Here is what the Bible tells us:

- God exists.

- God created the universe and controls every little event that takes place, right down to the atomic level.

- God provides order in our universe and therefore God defines our reality.

- God knows all, including the past, the present, and the future.

- God is all powerful and can make his ghostly little electrons and photons do anything he desires.

For **from** him and **through** him and **for** him are all things. To him be the **glory** forever! Amen.

(Romans 11:36)

DIVE DEEP

Why do you think so many people have an easier time believing in billions of parallel universes instead of believing in God? How about you? Where do you place your faith?

In the final chapter of this book, I will describe what I believe explains all of these results. But for now, I would prefer to share some more findings about the building blocks of our world.

THE ILLUSION OF PHYSICAL REALITY

THE UNCERTAINTY PRINCIPLE

In 1927, Werner Heisenberg published his paper on uncertainty. Like Einstein, Heisenberg proposed something that is not intuitive at all. His theory relates to the position and velocity of particles. It states that the more precisely the position of a particle is known, the less precisely the velocity can be known, and vice versa. Stated in other words, if you know exactly where a particle is, you can't know how fast it's moving or where it's going. If you know exactly how fast a particle is moving, then you can't know where it is. This theory states that our inability to know both position and velocity at the same time isn't because we don't have the capability to measure them both, but because of a physical law—we just aren't *allowed* to know both at the same time, or maybe both don't exist at the same time.

I don't know about you, but this simple little theory rubbed me the wrong way. For years I had known that we can easily know the position and velocity of something like a billiard ball, and therefore could predict where it would be in the near future. It is this simple understanding that has allowed us to shoot rockets to the moon and cannonballs to enemy targets. But with the uncertainty principle, there is no certainty about where a particle will be in the future. Therefore, the future is completely uncertain.

Again, nobody knows why this law doesn't apply in the macro world, but since 1927 we have known it to be true for the quantum world.

One implication of this principle is that the location of a particle which is fully at rest, with zero velocity, can't be nailed down. Particles at rest seem to appear in one particular spot and then suddenly appear in another. They don't move there; they just reappear in the new location. So, the particle has no velocity, but its location is uncertain at any particular instant.

Once again, I found myself as disturbed by science as I was by the Bible. In fact, as I learned about the uncertainty of a particle's rest position, I couldn't help but see a similarity to the Biblical writings about mountains that could jump into the sea. What if every little pesky particle that makes up a mountain suddenly appeared in a different spot? The mountain would have "jumped" from one location to another.

Is this possible? Yes, but scientists will tell you that it is very, very improbable.

Presumably, the stationary particles in the mountain are jumping around because of uncertainty, but always in different directions so that the sum total of their "jumping" leaves the mountain in the same place. Scientists will tell you that it is possible for each particle to jump in the same direction at the same time, which would cause the entire mountain to move. Weird stuff—and similar to the weird stuff described by Jesus.

It was at this point in my journey that all of this information was becoming a bit too much for me to ignore. The Spirit was starting to work on me not through my weakness, but through my strength—my cautious and analytical mind and the science that I had put so much faith in.

Like the results from the double-slit experiment, the uncertainty principle has been tested and examined over and over, and it always comes out the same. In fact, this principle is used to initiate nuclear reactions and in other high-tech engineering projects.

Might this be part of the veil that separates us from fully knowing God? It seems very weird to me that we aren't allowed to know these two simple things—location and velocity—at the same time, other than that it would allow us to predict some future events.

Hmm.

A TEN-DIMENSIONAL WORLD

All of the facts that I have shared with you up to this point make up what is called quantum mechanics. Even though we don't understand what all of this means, it hasn't stopped engineers from applying what we have learned. Lasers, computers, and microwave ovens are three examples of products that were designed using quantum mechanics.

THE ILLUSION OF PHYSICAL REALITY

The equations that physicists use to describe quantum behavior and that allow engineers to design products based on quantum behavior are ten dimensional.

Here we go again! What does this mean?

Many of you will remember studying algebra in school, and whether you liked it or not, you may recall dealing with two or three variables at a time. The rules for manipulating algebraic equations with two or three variables are the same rules used for equations with any number of variables. The equations become more cumbersome, but there is no difference between equations that handle two variables or ten.

He is before all things, and in him all things hold together. (Colossians 1:17)

Often, these polynomial equations are used to describe our physical environment, where x, y, and z represent our three-dimensional space—length, width, and depth. A fourth dimension, t, is often added to represent time. So, the world we experience on a daily basis can be described mathematically in four dimensions (length, width, depth, and time). The math required for several more dimensions is easy to do, but what would the extra dimensions represent?

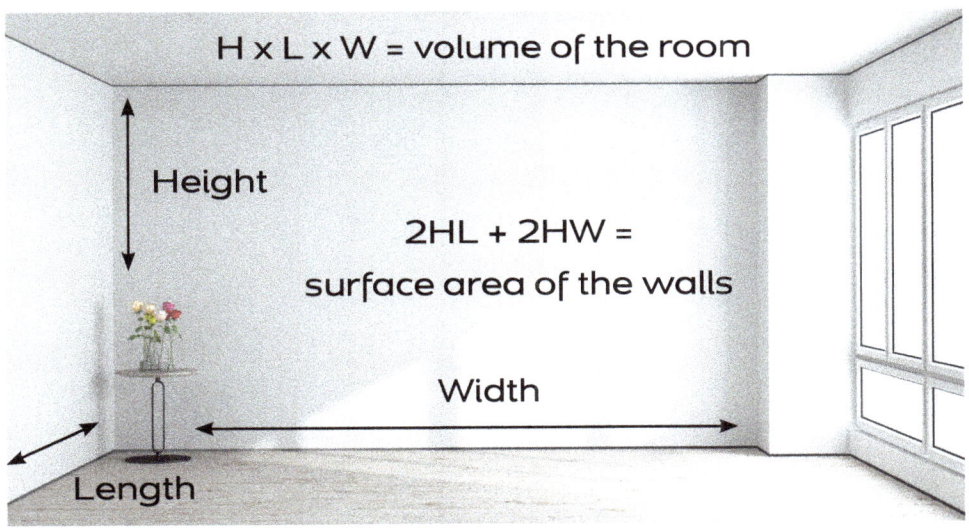

These are the 3D equations with simple physical meaning

As mentioned, the math that describes the quantum world only works with ten dimensions. Sometimes it requires 11 dimensions and this extra dimension is a 2nd dimension of time...so time is 2 dimensional in these equations. Nobody knows what this means, but this kind of math certainly describes how the quantum world operates. Once again, this is a little unsettling.

Let's think about what implications might come from a ten-dimensional world. If God's creation is ten-dimensional—whatever that means—then God must be at least ten-dimensional. As I learned this, I couldn't help but reflect on Scripture that says we can never fully understand God, just as we can't fully understand what dimensions five through ten represent.

Let's assume for a moment that God is a ten-dimensional being, and that we don't have the capacity to understand more than the four dimensions of our classical world. How might God choose to communicate with us? Let me share an illustration with you.

Let's pretend that there are two-dimensional beings such as those described in the book *Flatland: A Romance of Many Dimensions* by Edwin A. Abbott.

These two-dimensional beings could live in a single plane, like a windowpane. Let's pretend that you decided to visit the flatlanders. As a three-dimensional being, it would be quite difficult to get the flatlanders to understand what you are really like. If you were to enter their flat world, they would only see intersections of you. Your fingers would appear to be two-dimensional circular cross-sections of fingers. Your stomach and head would be bigger circular intersections. The flatlanders would likely think that you are just a bunch of circles and ovals that seem to change without any rhyme or reason.

So how might you devise a plan to get the flatlanders to understand who you are and what you look like? The best way might be to create a two-dimensional being who knows and understands you and then send this two-dimensional person into the flatlanders' world. This two-dimensional being would still be greatly challenged to get the other two-dimensional people to understand what you, a three-dimensional being, are really like, but at least he or she would be able to communicate with them on their terms rather than just confuse them with weird intersections.

THE ILLUSION OF PHYSICAL REALITY

Might God, presumably a greater than ten-dimensional being, create a four-dimensional being in his son Jesus and ask him to do his best to describe to us who God is?

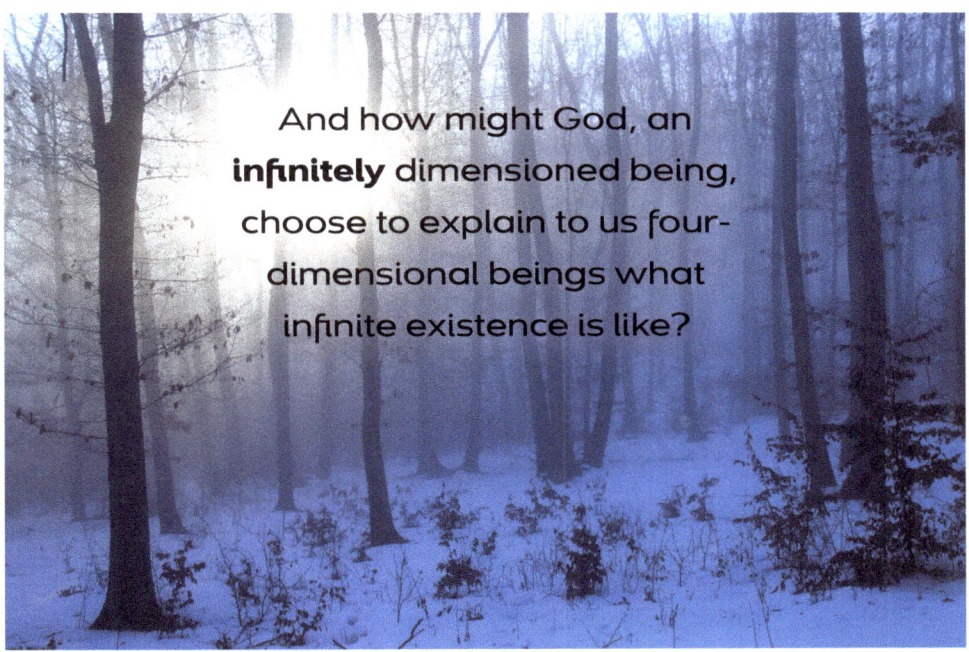

The Son is the image of the invisible God, the firstborn over all creation. (Colossians 1:15)

No one has ever seen God, but the one and only Son, who is himself God and is in closest relationship with the Father, has made him known. (John 1:18)

I know that Jesus ultimately came to earth to save our souls, but while he was here, he described to us as best he could, in our four-dimensional reality, what our Father in heaven is really like. We often get confused when we try to understand God, but if he is at least ten-dimensional, it is no wonder! And again, doesn't the Bible tell us that we can't fully understand God, and if we could we wouldn't be able to handle it?

Am I saying that I believe God is ten-dimensional and that Jesus, a four-dimensional being, was sent to describe him to us? Not necessarily, but this thinking does make me want to quit trying to second-guess God because he knows things that I will never be able to know or understand.

It makes me more willing to accept that the things I can't make sense of, make sense to God.

It makes me ask, "Who am I, a clueless four-dimensional being, to question Almighty God?!"

It makes me want to learn as much as I can about God from his word in the Bible, his word in the Flesh, and his spirit in the Holy Ghost.

Bottom line for me is that in the eternal kingdom of heaven, God knows all and is in total control. God is sovereign. Things in our limited understanding will often not make sense to us, but just like a child who doesn't yet understand the ways of the world, we need to accept that our Father loves us and will ultimately do what is best for us—in the eternal sense, not necessarily in the worldly sense.

So, we may get sick, or suffer due to injustice, or die at the hands of evil, or fail where it is obvious that we should have succeeded. But who are we as mere four-dimensional beings to question Almighty God? I think it makes sense that we are called just to be obedient to God. Anything else would in effect be saying that we know better than God, or in some very dangerous way that we think we are God. Isn't this what happened to the angel Lucifer?

SCHRÖDINGER'S BOTHERSOME CAT

I've thrown a lot at you, but stick with me for one last piece of information before I attempt to wrap it all up in the final chapter.

In 1935, Erwin Schrödinger sent Albert Einstein a letter describing a thought experiment that was meant to discredit all that had been found in the double-slit experiment. (Thought experiments have been a key part of the discovery process throughout history partly because they require no setup, no apparatus, and no funding, but also because they depend on imagination and intelligence and there are some people in this world who enjoy using both.)

THE ILLUSION OF PHYSICAL REALITY

Schrödinger's thought experiment goes something like this:

- Imagine a cat placed in a box.

- Also placed in the box is a radioactive atom that can decay at any moment.

- Also, in the box near the cat is a Geiger counter, which detects alpha particles that have been spit out by the decaying atom.

- Attached to the counter is a diabolical mechanism that releases poison into the air if an alpha particle is detected (obviously Erwin wasn't much of a cat lover).

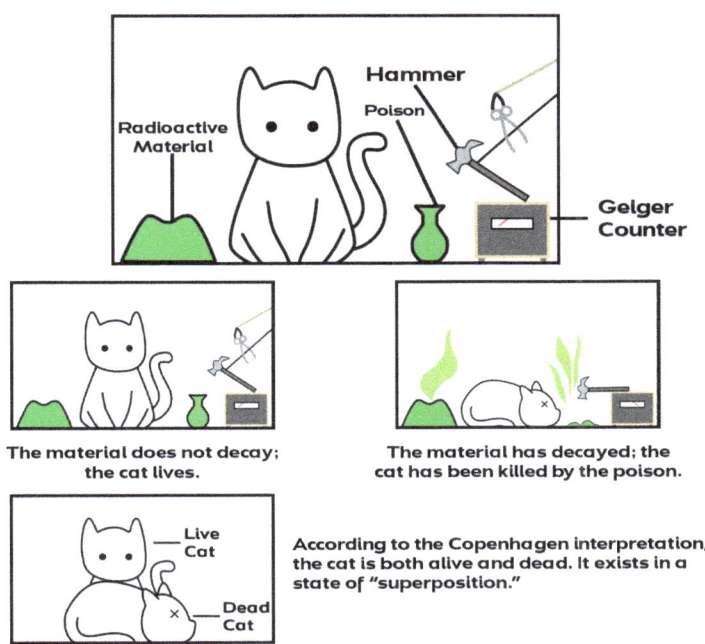

Now everything is ready to go in our thought experiment. The decaying atom is left in the box with the cat, Geiger counter, and poison just long enough to have a 50 percent chance of decaying and spitting out an alpha particle. Then, the radioactive atom is removed from the box without opening it and observing the cat.

According to classical physics, the decaying atom either spit out an alpha particle or didn't. But according to quantum theory, it both spit out the alpha particle AND it didn't. The actual fate of the cat wouldn't be determined until someone took a look into the box to see what had happened. Prior to observation from a person, the atom only existed in its wave function, which means that it would have existed in a superposition of states—both having spit out the radioactive alpha particle and not having spit it out.

So, the Geiger counter would have detected the alpha particle and not detected it. And the poison would have been released and not released. And the cat would be both alive and dead, and neither alive nor dead, until someone looks into the box, which would collapse the alpha particle's wave function into reality—into being a particle that either spit out the radioactive alpha particle or didn't spit it out, but either way, that activity would have happened in the past, even though it was not determined until the moment someone looked into the box.

This thought experiment was Schrödinger's way of bringing the craziness of the quantum world into the normal world. It was his attempt to show that this idea of ghostly superposition of states was ludicrous. Einstein tended to agree with Schrödinger. While neither was able to find any flaw in the quantum thinking of the day, both thought it was crazy to think that a cat could be both alive and dead at the same time—the animal version of ghostly superposition.

THE ILLUSION OF PHYSICAL REALITY

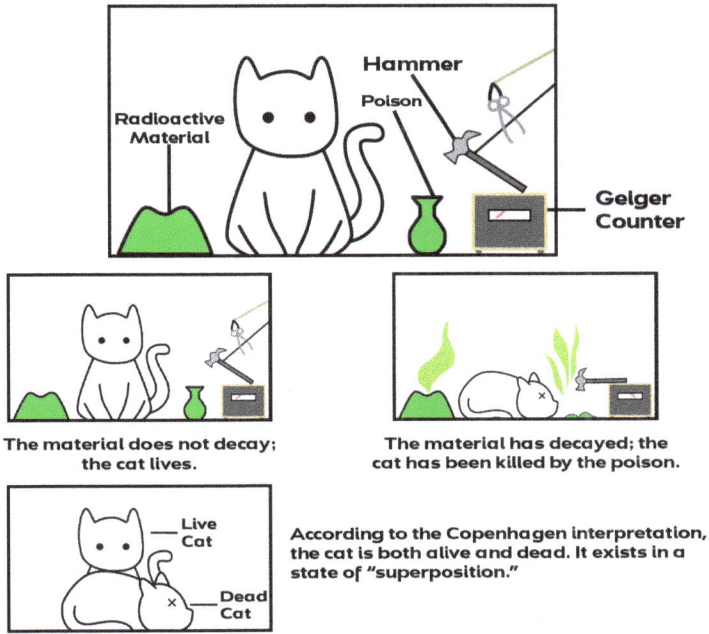

According to the Copenhagen Interpretation, the cat IS both alive and dead. It exists in a "superposition of states" until the top is opened and someone observes the "reality," which THEN comes with a history.

In the 1920s the world's greatest physicists gathered to deal with this thought experiment. After much discussion and debate, their conclusion, deemed the Copenhagen Interpretation, stated that the cat is both dead and alive until someone opens the box to look. Then, like the alpha particle, one of those realities collapses into being, and the cat is either only dead or only alive, not both.

To this day, this is one of the most popular interpretations by physicists of this thought experiment. Schrödinger, the author of this thought experiment, was later quoted as saying, "I don't like it and I'm sorry I ever had anything to do with it."

Physicists who are so often thought to be boring and even somewhat nerdy—definitely not the wild type—believe this radical view of our world! Really?

Wow!

But wait, there's more . . .

Now, let's assume that there is enough food, water, and room for the cat to live happily for one year before the box is opened. Of course, whether the cat ends up being dead or alive, one thing is for sure: That box is going to stink!!

So, as most physicists believe, the cat would be exactly as quantum mechanics describes it: both dead and alive and neither dead nor alive for a full year. But upon opening the box, the wave state of the particle would collapse, and the cat would be dead or alive—and it would be such with a history!

I know this sounds crazy. Really, I do! I have read several books on the subject and visited several university science department websites trying to disprove all of this craziness (and I invite you to do the same). But alas, of those authors and professors who've written about it, none of them had any other explanation than that this is how it all works. I believe physicists are unsettled by this, but they are okay with being unsettled. The fact is, they believe the cat wouldn't be dead or alive until the box is opened. It would be both . . . and neither.

This little thought experiment, which has become every bit as famous (in small circles of course) as the double-slit experiment, brings up another new concept not yet discussed. As I previously mentioned, when the box is opened, the fate of the cat is determined—and that determination comes with a one-year history.

WOW AGAIN!!

What does this mean? Can something come into reality a minute ago with a one-year history? Once again, welcome to the wonderful world of quantum physics! This is exactly what quantum physicists believe.

(And by the way, if you're wondering if the cat could collapse the wave, the belief is that only a person can do this. It requires a being with a conscious mind—someone who can think about their thinking.)

Upon learning all this, I couldn't help but think about God and creation. Might the creator of the universe in all his power be able to create a universe with a history?

THE ILLUSION OF PHYSICAL REALITY

You betcha!

Might our universe have been created 10,000 years ago, but have a 15-billion-year history?

You betcha again!

Even atheist physicists would have to agree that, based on the Schrödinger's cat interpretation, our 13.7-billion-year-old universe didn't really exist until it was observed. And at that point it not only existed but did so with a very long history.

If the world didn't collapse into reality until someone observed it, might it have done so with a very long history?

Interesting, right?

Do I believe this interpretation of Schrödinger's thought experiment explains all of the controversy regarding the age of our universe? Not necessarily, but just the possibility of it makes me think it doesn't matter what I think about the age of the universe. God can do whatever he wants, regardless of what we choose to believe.

I believe that our concept of time, which I previously shared is very limited and most likely flawed, causes us to have difficulty fully understanding things that depend on time. I also think our flawed understanding of time serves to weaken our faith in God. So now I choose to believe that I don't understand anything, and God knows all. Therefore, who am I to question him?

It's time for me to take several positions about what I believe, which I will do in the final chapter.

WHAT IT ALL MEANS

I've shared a lot of information with you, and much of that information is strange and mind-boggling. So, I can imagine that over the next days, weeks, and months you'll try to make your own sense of it.

I wrestled with most of what I've shared with you for about five years before I finally made my own sense of it. This final chapter describes the sense I have made of all of these strange scientific facts and theories. It might also work for you.

First, I'd like to start with a devotion from Sarah Young's *Jesus Calling: Enjoying Peace in His Presence:*

> May 21
>
> I, THE CREATOR OF THE UNIVERSE, am with you and for you. What more could you need? When you feel some lack, it is because you are not connecting with Me at a deep level. I offer abundant Life; your part is to trust Me, refusing to worry about anything.
>
> It is not so much adverse events that make you anxious as it is your thoughts about those events. Your mind engages in efforts to take control of a situation, to bring about the result you desire. Your thoughts close in on the problem like ravenous wolves. Determined to make things go your way, you forget that I am in charge of life. The only remedy is to switch your focus from the problem to My Presence. Stop all your striving and watch to see what I will do. *I am the Lord!*

What, then, shall we say in response to these things? If God is for us, who can be against us? He who did not spare his own Son, but gave him up for us all—how will he not also, along with him, graciously give us all things? (Romans 8:31-32)

But as for me, I watch in hope for the LORD, I wait for God my Savior; my God will hear me. (Micah 7:7)

WHAT IS THE TRUE REALITY?

In the preceding chapters, you learned that scientists—and the rest of us—are pretty clueless as to the true reality of this existence. Certainly, the reality of the world isn't what it seems to be from our very limited vantage point.

Some scientists believe that reality is created by our beliefs. This is known as the "placebo effect".

Does what we believe really affect our physical reality?

Remember that what we observe causes the ghostly wave functions to collapse into reality? Remember Einstein's quote about the moon really being there whether we're looking at it or not?

These are unsettling thoughts about what reality really is.

One of the prominent theories that can explain much of the weirdness we've been exploring is that there are an infinite number of universes—a new one created each time anyone observes something, or chooses something, or believes something.

After several years of study and reflection, I came to a fork in the road, where taking one path would mean believing that I create my own reality through my beliefs, and taking the other path would mean believing that there is only one reality, like most of us believed prior to learning about this new science of the past 100 years. However, for this to be true, this one-reality path would have to somehow explain all of the weirdness found in science.

WHAT IT ALL MEANS

John Gribbin, the scientist who wrote *Schrödinger's Kittens and the Search for Reality*, said this in his book:

Schrödinger's Kittens and the Search for Reality
By John Gribbin

"For the Universe to exist as one reality, not a superposition of states, the Copenhagen Interpretation strictly speaking requires the existence of an observer outside the Universe to do the collapsing of the wave functions. So, some cosmologists have turned to the many-worlds interpretation, preferring to argue that there really are many universes, each occupying its own region of space and time, and all tracing their origins back to the Big Bang." (Page 163)

This describes the fork in the road that I experienced. Either there is "the existence of an observer outside the Universe to do the collapsing of the wave functions," i.e., God, or there are "many worlds."

Neither is provable. And beliefs that we cannot prove are what I call faith.

When contemplating this fork in the road for some time, here is what filled my mind:

On the one hand, if I knew that what I choose to believe creates my reality, would I choose a reality based on a belief in Christ? Did I believe that this version of faith would serve me well even if it were only my own personal reality? I decided it would, so choosing to believe in Christ seemed like a wise choice if reality is created by my beliefs.

If, on the other hand, I choose to believe that God created everything and continues today to collapse the wave functions into reality as John Gribbin says, then I would be believing in one reality, not in an infinite number of realities that are continuing to branch off from each other at every moment. This belief can explain all the weirdness, as long as we also choose to believe God is . . .

- creating the world constantly (as the alpha and the omega), and
- omnipresent—everywhere at once (just like light), and
- omniscient—able to know everything we're up to and everything we know, and
- omnipotent—having the power to manipulate all matter and forces at will.

WHAT IT ALL MEANS

Two Paths
(my experience in mid-90s)

No objective reality. We create our own reality. Infinite number of universes.

The great I AM controls everything.

Path A Path B

Either way, I live by **Faith**.

He is before all things, and in him all things hold together. (Colossians 1:17)

At this point in my life, when I was 38 or so, I realized that this contemplation had become an academic exercise—either way I would choose to believe in Jesus and in the God of the Bible. I would choose to believe that Christ is Lord and is alive and gives life in the here and now and also in the everlasting.

That didn't bring me to my knees, though. But it opened the door for that.

The Holy Spirit was also at work in my heart through another book I had read and through some other things that had happened in my life. Then one night while alone and flipping through channels on the TV, I came across a televangelist. I couldn't tell you who he was or what show it was. But I *can* tell you that he would normally NOT have grabbed my attention because he was a caricature of a televangelist: his hair, the way he spoke, and even his style of preaching.

But he asked this question in the minute or two that I was watching: "Do you believe Jesus is the risen Christ? Do you believe he died for your sins?"

I answered yes. I realized that the time was now. So, I turned the TV off and, sitting in my reclining chair, I prayed the sinner's prayer. I confessed to Jesus that I am a sinner and I needed his grace to save me. I thanked Jesus for what he did for me on the cross. I thanked him for his patience with me. I didn't understand how this would change my life, but I was finally ready to find out.

There . . . it was done. No more running. No more trying to figure out how this reality happened without a creator.

I began to find a new kind of peace even though I had already been a pretty peace-filled guy.

It's been a fun journey ever since that February night in 1995. And ever since I made this quantum leap of faith, I see God's presence and God's miracles all around me.

"Faith is to believe what we do not see, and the reward of faith is to see what we believe." (Saint Augustine)

SO HOW SHALL WE LIVE?

I have come to believe these things:

- God created the universe and controls every little quantum event that creates the foundation for all life and reality, thus constantly providing order in the midst of chaos (so creation wasn't a one-time thing "in the beginning").

- God knows all things in all dimensions, including the past, present, and future.

- God is all powerful and can make his ghostly little electrons and photons do anything he desires.

God really does, as the children's song says, "hold the whole world in his hands" and more!

Praise God!

For me, believing in God as the Father, the Son, and the Holy Spirit is life; the rest is just details.

Let me share just a few more thoughts about some of those details.

MIRACLES

Quantum weirdness has taught me to believe what one author suggested many years ago. (Unfortunately, I don't remember the author's name.) He said that no biblical miracle is impossible or breaks any laws of physics. It has taken me several years to come to believe this, but I do.

> "There is no Biblical miracle that breaks the laws of the new physics."

While no miracle breaks the laws of quantum physics, biblical miracles are highly, highly unlikely. So, while I think that no miracle is supernatural (depending on one's definition of supernatural), I also think that all things, because of God's constant control and re-creation and giving of order, are supernatural. How's that for duality?

So, believing that miracles are just a matter of highly unlikely things happening, I started to notice all kinds of unusual occurrences happening around me.

Let me explain this a bit further using the flip of a coin. If I were to flip a coin right now, would it be a miracle for it to come up heads?

How about five heads in a row?

How about twenty? One hundred? One thousand?

Do you see my point? No single flip would be a miracle in itself, but put them all together and something "supernatural" must be going on.

How is it that I just happened to bump into that person right when I needed to? Was it perhaps a miracle? How is it that the money I needed for that emergency showed up at just the right time?

Might miracles be all around us constantly, orchestrated by an almighty and all-loving God?

LACK OF FAITH

How many of my problems relate to my fears? God has promised to love me, protect me, provide for me, and more. I am told to "fear not" (with the exception of fearing God of course). What problems could I have if I truly did this?

Who are you that you fear mere mortals, human beings who are but grass, that you forget the LORD your Maker, who stretches out the heavens and who lays the foundations of the earth, that you live in constant terror every day because of the wrath of the oppressor, who is bent on destruction? (Isaiah 51:12b-13a)

Have you ever met someone with this "fear-not" kind of faith? How amazing would it be to be that person?

DIVE DEEP

How many of your big problems are rooted in lack of faith? How might you start to change this?

By the way, I'm sooooo far from being where I want to be with my faith. It's easier to name it and say it than it is to do it. But I understand that the problem is my lack of faith, not that I might lose my job, or that I am dangerously ill, or that some people I love might not love me, or whatever hardship might come.

If I am filled with deep faith, I'll be just fine in the long run.

WHAT IT ALL MEANS

If I could be filled with deep faith, here's what I think it would look like:

- I would be free from fear because nothing and no one can separate me from my deep faith or from God's amazing love and promises. Paul of Tarsus seemed to have this kind of faith.

- I would see God's love for me and trust that he will take care of me while I also work to take care of myself and others—yet another duality in my beliefs.

- I would realize that this earthly life is a journey, and while I'll never be fully where I desire to be in my faith, I can work to make tomorrow more faith-filled than today.

- I would experience more of the abundant life that Jesus promised as I grow—the life based on the eternal things I need, not necessarily on the earthly things I want. By the way, do you think the early martyrs lived an abundant life? I do. And I think we see this when Stephen gets stoned to death. (You can read about Stephen's death in Acts 7:54-60.) Notice his reaction to one of the worst earthly events that could happen to a person. This is what real faith looks like.

The righteous perish, and no one takes it to heart; the devout are taken away, and no one understands that the righteous are taken away to be spared from evil. (Isaiah 57:1)

For we know that if the earthly tent we live in is destroyed, we have a building from God, an eternal house in heaven, not built by human hands. (2 Corinthians 5:1)

So, I am trying to live by faith in God—the Father, Son, and Holy Spirit. This faith is based on what the Bible, the Holy Spirit, and fellow Christians tell me, not what my experience and feelings in this physical world tell me.

This is the "faith of a child," and I often don't do too well with it. My job is simple to understand but hard to do—discern God's will, and then do it. Not tweak it, not debate it, not question it, just do it. He knows all. He promised to care for me. Who am I to doubt this and question his wisdom?

Where were you when I laid the earth's foundation? Tell me, if you understand. (Job 38:4)

Do you know the laws of the heavens? Can you set up God's dominion over the earth? (Job 38:33)

In an eager attempt to understand God's will, I spend significant time in the Bible and in prayer, including listening prayer, and also in Christian community. I try to pay attention to the miracles that happen around me so that I might understand what God is up to and how I might be able to join him.

I also sin, and get lazy, and fail miserably. And confess all this and start again. It's a journey, after all—a good one!

I hope my journey, my thoughts, and the facts I've shared about some of the new science have been a blessing to you. I know that pulling them together and writing them down has been a blessing to me.

If you have any questions, feel free to contact me at: faithandphysics@gmail.com.

God bless you on your journey!

ACKNOWLEDGMENTS

God has brought many people into my life who have challenged, encouraged, and supported me. First and foremost, I would like to acknowledge my wife, Amanda, for her love and encouragement in our almost 50-year journey together.

Secondly, thanks to my brother Ron, for his persistent encouragement to write down my journey and his help in publishing it. With his support, it can get in more hands to encourage more people.

Next, thanks to the Reformed Church in America for allowing me and Meredith Nieuwsma to work on the initial version of my story so participants in my class could have a takeaway, which has become this book.

I would also like to acknowledge Maryanna Young, Heather Goetter, and the rest of the team at Aloha Publishing for their tireless work in bringing this book to life.

Lastly, I want to acknowledge the many people who shared their appreciation for what I learned on my way to the cross. Their sincere feedback gave me the courage to continue to share with even more people.

ABOUT THE AUTHOR

Rodger Price lives in Holland, Michigan with his wife, Amanda, who he has been married to for over 4 decades. He loves meaningful conversations and training leaders. He believes that leadership skills are not always inherent, but when those skills are offered through experiential and reinforced learning models, they can be developed. His work focuses on the belief that each person has a unique design for their life.

He began his career as an automotive engineer and, after considering making a transition to teach high school physics, transitioned into being a leader development expert. In 2014, he founded Leading by DESIGN and is currently the managing partner.

A longtime agnostic, Rodger started exploring 20th century physics in his mid-thirties which led to insights about how much faith it takes to not believe in God. Uncovering the power of physics led him to faith in Christ which changed how he lives his life every day.

Rodger enjoys hiking, playing tennis, spending time with family, and training leaders with unique hands-on learning. He also loves to stay current on the latest information about physics and is passionate about sharing how understanding science opens people's understanding of faith.

CONNECT WITH ME

I would love to connect with you.
Please email me: faithandphysics@gmail.com

You can also connect with me @Linkedin
linkedin.com/company/faith-and-physics-uncovered

www.ingramcontent.com/pod-product-compliance
Lightning Source LLC
Chambersburg PA
CBHW040002080526
44586CB00027B/2848